SEWERAGE AND SEWAGE DISPOSAL

BY
ALLEN WISDOM, SOLICITOR

Published by
BARRY ROSE (PUBLISHERS) LTD.

Printed by
T. J. Hunt Ltd, London SE15

CONTENTS

FOREWORD

For those whose work brings them into contact with sewers and sewage disposal works, Part II of the Public Health Act, 1936, provides the main legislative framework. Part II has to some extent suffered the ravages of time, as well as being amended by the Water Act, 1973, when sewerage and sewage disposal was transferred from the local authorities to the water authorities. This booklet provides an updated version of those sections in Part II (with references to other sections of the 1936 Act and later public health legislation where necessary) which affect sewers and disposal works, together with some of the case-law which has grown up around the subject-matter.

ABBREVIATIONS

PHA, 1936	Public Health Act, 1936
PHA, 1961	Public Health Act, 1961
WA, 1973	Water Act, 1973
CPA, 1974	Control of Pollution Act 1974

1. DUTY TO PROVIDE SEWERAGE AND SEWAGE DISPOSAL FACILITIES

(WA, 1973, s.14)

Each water authority is under the statutory duty to:—

> "provide either inside or outside its area such public sewers as may be necessary for effectually draining its area and to make such provision whether inside or outside its area by means of sewage disposal works or otherwise for effectually dealing with the contents of its sewers."

Up to April 1, 1974, this function had been carried out by the local authorities and the effect of s.14 of the Water Act, 1973, was that on and after that date the function passed into the hands of the water authorities, and local authorities ceased to have any duties or responsibilities for sewerage and sewage disposal, subject to their agency arrangements for sewerage functions (referred to in para.2), and apart from isolated instances where the local authority retains some duty or shares a power with the water authority in relation to sewerage.

As a consequence of this transfer of functions to the water authorities, s. 14 went on to provide that the requisite public health legislation should be exercised by water authorities and that references in such legislation to a local authority and its district should be construed respectively as references to a water authority and its area. The legislation comprises:—

Public Health Act, 1936, ss.15, 17-24, 27, 29-31, 33, 34, 36 and 42 Part XII, so much as relates to the above sections.

Public Health (Drainage of Trade Premises) Act, 1937

Local Government (Miscellaneous Provisions) Act, 1953, s.13

Public Health Act, 1961, ss.12-14, Part V and sch.2

and this has been amended and altered by s.14, sch.8, paras. 32-42 and sch.9 to the Water Act, 1973.

2. LOCAL AUTHORITIES AS AGENTS FOR WATER AUTHORITIES REGARDING SEWERAGE

(WA, 1973, s.15)

Arrangements normally exist between the water authority and local authorities (which could be a district council, London borough council or a new town development corporation) whose area is wholly or partly situated in the water authority area whereby the local authority discharges within its area the sewerage functions of the water authority other than sewage disposal and the maintenance and operation of sewers vested in joint sewerage boards or the Greater London Council immediately before 1st April, 1974. The arrangements cover the following matters:—

(1) The local authority has to prepare and carry out an annual programme for the discharge of sewerage functions as approved by the water authority;

(2) Public sewers provided by the local authority are vested in the water authority;
(3) The local authority provides the necessary vehicles and equipment for maintaining the sewers;
(4) The water authority reimbursing the local authority for the cost incurred in discharging the sewerage functions;
(5) The local authority conducts prosecutions and other legal proceedings in connection with the discharge of the functions;
(6) In the event of the arrangements being ended the local authority officers to be transferred to the water authority and compensated for loss due to any variation or ending of the arrangements.

In the first instance, the arrangements were agreed between the water authorities and the local authorities prior to or around the time when the Water Act, 1973, became operative (April 1st, 1974) and, if not agreed, the terms were settled and finalised by the Secretary of State for the Environment. Occasionally, the water authority and the local authority failed to agree and if the two parties concluded that it was inexpedient in the interests of efficiency to enter into any arrangements, the matter was dropped and sometimes it occurs that a local authority do not operate any agency sewerage functions.

Existing arrangements can be varied or ended by the parties by agreement or in default the Secretary of State may do so.

Subject to the terms of the arrangements, the local authority may discharge the sewerage functions by a committee or sub-committee or an officer.

3. PROVISION OF PUBLIC SEWERS AND SEWAGE DISPOSAL WORKS

(PHA, 1936, ss.15; WA, 1973, ss.14, 40(2), sch.8, para.32)

The water authority is empowered either within or outside its area to construct public sewers and sewage disposal works. A sewer may be constructed in, under or over a street, or under a cellar or vault below a street, in which case the provisions of the PHA, 1936, ss.279, 282, the Public Utilities Street Works Act, 1950, and the Water Act, 1945, sch.3, Part VI relating to the breaking up of streets, must be complied with: or a sewer may be constructed in, on or over any land not forming part of a street – this includes a building which may have to be demolished (*Hutton v Esher UDC* [1973] 2 All E.R. 1123) – after giving reasonable notice to every owner and occupier of the land. The land comprising the sewer route does not have to be purchased – reasonable notice is given under s.15, notice of entry is served under s.287, the sewer after construction vests in the water authority under s.20, and the owner or occupier can claim compensation under s.278. Laying the sewer may be done by the water authority, or by the local authority under the sewerage agency arrangements (see para.2).

Before constructing a public sewer in the area of a local authority the water authority is required to consult the local authority and inform it of the construction of the sewer. In addition before adopting plans for sewering any part of its area, the water authority has to notify its proposals to each parish council or meeting to be served by the works. The water authority, before constructing a sewer which will cross or interfere with a watercourse or works vested in or under the control of a land drainage authority (i.e. a water authority or internal drainage board), must give notice of its proposals to that authority before adopting plans for the sewer construction; if the drainage authority object to the proposals by serving notice on the water authority within 28 days, the water authority cannot proceed unless the objection is withdrawn or the Secretary of State after a local inquiry has approved the proposals with or without modification.

The water authority may construct sewage disposal works on any land acquired or lawfully appropriated for the purpose – this includes extending an existing works, and may by agreement acquire, whether by way of purchase, lease or otherwise, any sewer or sewage disposal works, or the right to use any such sewer or works. In pursuance of its sewerage and sewage disposal functions a water authority may compulsorily acquire land under s.306 of the Public Health Act, 1936, which section is preserved under the Control of Pollution Act, 1974, s.108(1), sch.3, para.27 and the Control of Pollution Act, 1974 (Commencement No. 1) Order, 1974 (S.I. 1974, No. 2039 (c.33), sch. Part III). A purified effluent can be discharged to a watercourse (see p.16).

In the event of failure by the authority to discharge its duty to provide sewerage and sewage facilities, complaint should be made to the Secretary of State pursuant to ss.321 to 324 of the PHA, 1936 (see para. 30). It has been held that this duty will not be enforced by way of an action for mandamus (*Passmore v Oswaldtwistle UDC* [1898] A.C. 387), and an action for damages does not lie for failure to provide sufficient sewers (*Clark v Epsom RDC* [1929] 93 J.P. 67). If an authority is negligent in the exercise of its statutory powers, as in the construction of a sewer, so as to cause a nuisance and injury, then the injured party can bring an action (*Dent v Bournemouth Corpn*[[1897] 66 LJQB 395). There is also the right to claim compensation under s.278 of the PHA, 1936.

If a sewer is laid in a negligent manner the water authority may be liable and thus responsible for the acts of its servants and contractors (*Newsome v Darton UDC* [1938] 3 All E.R. 93, C.A.). An authority is not bound to provide a sewer for every house in its area (*Kinson Pottery Co v Poole Corpn* [1899] 2 Q.B. 41, D.C.).

4. SOME DEFINITIONS

(PHA, 1936 ss.90(4), 343)

Having set the scene, it is useful at this stage to refer to a number of statutory definitions and terms in common use; these include:–

SEWAGE DISPOSAL WORKS – not defined in the PHA, 1936, but a repealed Act referred to "works ... for the purposes of receiving, storing, disinfecting, distributing or otherwise disposing of sewage." The term includes machinery and equipment for the works and any necessary pumping stations and outfall pipes (PHA, 1936, s.90(4)).

PUBLIC SEWER – This is defined in para.9, and the term includes manholes, ventilating shafts, pumps and other accessories belonging to the sewer (PHA, 1936, s.90(4)).

CONTENTS OF SEWERS – This refers to (a) foul water from toilets and domestic baths and sinks, etc; (b) surface water from roofs and roads; but not (c) trade effluents, where the right of an occupier of trade premises to discharge trade effluent into the public sewer of the water authority is governed by the Public Health (Drainage of Trade Premises) Act, 1937, as amended.

DRAIN – means a drain used for the drainage of one building or of any buildings or yards appurtenant to buildings within the same curtilage (PHA, 1936, s.343(1)).

SEWER – does not include a drain (as defined above), but includes all sewers and drains used for the drainage of buildings and yards appurtenant to buildings (PHA, 1936, s.343(1)).

PRIVATE SEWER – means any sewer which is not a public sewer (PHA, 1936, s.343(1)).

PREMISES – includes messuages, buildings, lands, easements and hereitaments of any kind (PHA, 1936, s.243(1)).

SURFACE WATER – includes water from roofs (PHA, 1936, s.90(1)).

5. VESTING OF SEWERS AND SEWAGE DISPOSAL WORKS IN THE WATER AUTHORITY

Sewers can be vested in the water authority either by a vesting declaration under s.17 of the PHA, 1936, or in accordance with s.20 of that Act as replaced by the WA, 1973, s.40(2), sch.8, para.35. An agreement to adopt a sewer or sewage disposal works at a future date may be made under s.18 of the PHA, 1936. These provisions are dealt with in paras.6,7 and 9.

The effect of vesting a sewer in the water authority confers on the authority for controlling and maintaining it, an absolute property in the sewer and the space it occupies (*Bradford v Eastbourne Corpn* [1896] 2 Q.B. 205), although no interest in the soil surrounding the sewer vests in the authority (*Newcastle Corpn v Woolstanton Ltd* [1947] 1 All E.R. 218); the interest in the sewer appears to be a legal estate in fee simple determinable when the pipe or conduit ceases to be a sewer (*Tithe Redemption Comrs v Runcorn UDC* [1954] 1 All E.R. 653). Private rights enjoyed regarding a sewer before it became vested cannot be interfered with (*R v Staines L.B.* [1888] 60 L.T. 261), and a drain cannot be disconnected from a public sewer (*Eastwood Bros Ltd v Horley U.C.* [1901] 49 W.R. 308) – but see para.12.

6. ADOPTION OF SEWERS AND SEWAGE DISPOSAL WORKS

(PHA, 1936, s.17)

The water authority may at any time declare that a sewer or sewage disposal works (not being a sewer or works the construction of which was completed before 1st October, 1937) situate within its area or serving the area or any part of it shall become vested in the authority as from the date specified in the declaration. Notice of the proposal must be served by the water authority on the owner of the sewer or works and no further action may be taken until either two months has elapsed without an appeal having been lodged, or until any appeal has been determined. On the other hand, the owners of the sewer or works may request the water authority to make a vesting declaration.

An owner who is aggrieved by the proposal of the water authority to make a declaration may appeal to the Secretary of State within two months after notice of the proposal is served on him, and an owner aggrieved by the authority's refusal to make a declaration may appeal to the Secretary of State at any time after receiving notice of refusal, or if no notice is given him at any time within two months from making his application. The Secretary of State may allow or disallow the proposal or make a declaration and may if it thinks fit specify conditions, including the payment of compensation by the water authority and direct that the declaration it has made shall not take effect until specified conditions are accepted. In deciding whether a declaration should be made, the water authority, and the Secretary of State on appeal, shall have regard to all the circumstances of the case and in particular to the considerations specified in s.17(4).

A person who is entitled to use a sewer before a declaration is made, can use that sewer or any sewer substituted for it, to the same extent as if the declaration had not been made.

A water authority who is considering whether to make a declaration as to a sewer or sewage disposal works situate within the area of another water authority, or situate within its own area but serving another water authority's area, must give notice to the other water authority and cannot make a declaration until the other water authority have consented thereto or the Secretary of State on application to him has dispensed with the necessity for such consent either conditionally or unconditionally. When a water authority has made a declaration as to a sewer or sewage disposal works situate within the area of another water authority it must notify the fact to the other authority.

A vesting declaration cannot be made in respect of sewers or sewage disposal works vested in another water authority or a county council, board, British Railways Board or dock undertakers, except on their own application.

7. ADOPTION OF SEWER OR SEWAGE DISPOSAL WORKS AT FUTURE DATE

(PHA, 1936, s.18)

The water authority may agree with any person constructing or proposing to construct a sewer or sewage disposal works that should the sewer or works be constructed in accordance with the terms of the agreement the authority will on the completion of the work, or at some specified date, or on the happening of some future event, declare the sewer or works to be vested in it. The agreement is enforcable against the authority by the owner or occupier of any premises served by the sewer or works. This provision also applies to drains provided that the declaration is not made before the drain has become a sewer.

The water authority cannot make an agreement as to a sewer or drain or works situate within another water authority area until the other authority have consented to it or the Secretary of State has dispensed with the necessity of consent conditionally or unconditionally.

8. WATER AUTHORITY MAY REQUIRE PROPOSED SEWER TO BE SO CONSTRUCTED TO FORM PART OF GENERAL SYSTEM

(PHA, 1936, s.19; CPA, 1974, s.99, sch.2, para.6)

Where the water authority considers that a drain or sewer proposed to be constructed by a person is, or is likely to be, needed to form part of a general sewerage system which the authority has provided or proposes to provide, the authority may require him to construct the drain or sewer in a manner differing, as regards material or size of pipes, depth, fall, direction or outfall or otherwise from the manner in which he proposes, or could otherwise be required by the authority, to construct it. The person must comply with the water authority's requirements, but if aggrieved thereby he can within 28 days appeal to the Secretary of State who may disallow the requirements or allow them with or without modification.

The authority has to repay the person constructing the drain or sewer the extra expenses he reasonably incurs in complying with the requirements, and until the drain or sewer becomes a public sewer, the authority must also periodically repay him so much of any expenses he reasonably incurs in repairing or maintaining it as may be attributable to the requirements having been made and complied with. If any question arises as to the amount of payment to be made to the person this can be determined on his application by the magistrates' court or he may require this to be referred to arbitration – this will go to the Lands Tribunal.

Without prejudice to any other remedy, a person who constructs the drain or sewer otherwise than in accordance with the water authority's requirements is liable to a fine not exceeding £200.

This provision does not apply to a drain or sewer proposed to be constructed by railway or dock undertakers in or on land belonging to them and held or used by them for the purposes of their undertaking.

9. VESTING OF SEWERS AND SEWAGE DISPOSAL WORKS

(PHA, 1936, s.20, as substituted by the WA, 1973, s.40(2), sch.8, para.33)

In addition to the sewers and sewage disposal works vested in a water authority under ss.254 or 68 of the Local Government Act, 1972, as either section applies for the purposes of the WA, 1973, the following vest in a water authority:—

(a) all sewers and sewage disposal works constructed by the water authority at its expense, or vested in the authority pursuant to arrangements under s.15 of the WA, 1973, or otherwise acquired by the authority;

(b) all sewers constructed under Part IX of the Highways Act, 1959 (relating to the procedure for making up of private streets at the cost of the frontagers), except sewers belonging to a road maintained by the highway authority;

(c) all sewers and sewage disposal works with respect to which a declaration of vesting under Part II of the PHA, 1936, has taken effect.

Sewers which become vested in a water authority under the said ss.254 or 68 or this section are referred to as "public sewers".

10. AGREEMENT WITH COUNTY COUNCIL TO USE HIGHWAY DRAINS AND SEWERS FOR SANITARY PURPOSES, ETC

(PHA, 1936, s.21; WA, 1973, s.14(3))

A county council and the water authority may agree that drains or sewers vested in the county council as highway authority shall be used by the water authority for conveying surface water from premises or streets. This power can be exercised by local authorities (i.e. a district council or London borough council) as well as by water authorities.

And the county council and water authority may agree that public sewers vested in the water authority shall be used by the county council for conveying surface water from roads repairable by the county council.

In the event of a sewer or drain discharging directly or indirectly into the sewers or sewage disposal works of another water authority, the agreement cannot be made without the consent of that other authority in such terms as it thinks fit.

A county council or water authority must not unreasonably refuse to enter into an agreement or insist unreasonably upon terms unacceptable to the other party; nor can a water authority unreasonably refuse to consent to the making of an agreement or insist unreasonably upon terms unacceptable to either party thereto. Any question as to who is acting unreasonably is determined by the Secretary of State whose decision is final.

11. REQUISITIONING OF SEWERS FOR DOMESTIC PURPOSES

(WA, 1973, s.16)

This is a new provision somewhat similar to s.37 of the Water Act, 1945, which enables a landowner to require statutory water undertakers to lay mains for providing a supply of water for domestic purposes for new buildings. Section 16 provides that, subject to the appropriate conditions specified in s.16(2), (3) or (4) being satisfied, every water authority is under a duty to provide a public sewer for use for domestic purposes for draining premises (including new buildings proposed to be erected) in its area if so required by the owners or occupiers of the premises, or if the district council in whose area the premises are situated undertakes to meet the difference between one-eighth of the cost of providing the sewer and the drainage charges attributable to its use. The water authority may require a private landowner to deposit suitable security in the case of new buildings. If the conditions mentioned above are satisfied and the water authority fails to lay the sewer within six months or any longer period as agreed or determined, the authority becomes liable on summary conviction to a fine not exceeding £400, unless the failure was due to unavoidable accident or other unavoidable cause. A sewer is deemed to be used for domestic purposes if used for removing lavatory contents or removing cooking or washing water other than water used for the business of a laundry or that of preparing food or beverages for consumption otherwise than on the premises.

12. ALTERING OR CLOSING PUBLIC SEWERS

(PHA, 1936, s.22; WA, 1973, s.14(5))

The water authority may, as regards a public sewer vested in it, alter its size or course, or discontinue and prohibit its use, either entirely, or for foul water drainage, or surface water drainage. But a person who lawfully uses a sewer for any purpose cannot be deprived by the authority of its use for such purpose, until the authority (1) have provided a sewer equally effective for him for that purpose, and (2) have at the authority's expense carried out any necessary work to connect his drain or sewer with the sewer provided.

The water authority is under a duty to consult a local authority (ie. the district council or London borough council) before diverting or closing a public sewer in the area of the local authority, and also inform the local authority of the diversion or closure.

Sewers may be taken up, diverted or altered by British Railways Board, dock undertakers or land drainage authorities at their expense under s.330 of the PHA, 1936.

13. DUTY TO MAINTAIN PUBLIC SEWERS

(PHA, 1936, s.23)

It is the duty of the water authority to maintain, cleanse and empty all public sewers vested in it, subject to its right (described in para.14) in certain cases to recover its expenses in maintaining a length of public sewer.

Failure to discharge this duty can be enforced by way of complaint to the Secretary of State under ss.321-324 of the PHA, 1936. This duty does not extend to maintaining land or conditions outside the sewer, nor as regards obligations to third parties being strangers to the sewer (*Radstock Co-operative Socy v Norton-Radstock UDC* [1968] 2 All E.R. 59, C.A.).

14. RECOVERING COST OF MAINTAINING LENGTHS OF PUBLIC SEWER

(PHA, 1936, s.24; WA, 1973, s.40(2), sch.8, para.34)

In certain special cases the water authority can reimburse itself the cost of maintaining a length of public sewer vested in it. The owners of premises served by the length of public sewer in question are liable to pay the cost reasonably incurred by the water authority in maintaining that length. The cost is that proportion which the authority deem it fair to fix, having regard to all the circumstances of the case, including the benefit derived from the sewer length, the distance laid in each ownership, the point at which any work was necessary and the responsibility for any act or default rendering the work necessary. "Maintenance" includes repair, renewal and such improvement necessary for making the sewer length adequate for draining the premises served by it immediately before the improvement was undertaken.

Unless it is certified that immediate action is necessary, the water authority is required, not less than seven days before commencing the work to notify the owners of the premises known to it to be served by the sewer length of the proposed work and consider any representations made by any owner within seven days of service of the notice as to the need for and reasonableness of the proposed work. Either the owner concerned or the water authority may apply to the magistrates court to determine certain specified points. There is a restriction on the cost to be recovered when the water authority wish only to improve or enlarge the sewer length for serving additional premises.

This provision applies only to a length of public sewer under the following circumstances:—

(a) a length for the maintenance of which persons other than the local authority were immediately before October 1, 1937, responsible under some enactment or statutory scheme or order relating to combined drains, or an agreement, whereby the authority was entitled to require those persons to maintain that length of the sewer, or to abate any nuisance therein, or to contribute in proportions to, or indemnify the authority against, any expense incurred by the authority in maintaining it; or

(b) a length which was vested in the local authority immediately before October 1, 1937, but was not constructed at its expense or that of their predecessors and which lies in a garden, court or yard belonging to any of the premises served by the sewer or common to any two or more of them, or lies under a building comprised in any of those premises or lies in a roadway, footway, passage or alley used solely or mainly as a means of access to those premises or any of them, but is not a highway maintainable at public expense.

15. PROHIBITION ON ERECTING BUILDING OVER SEWER OR DRAIN

(PHA, 1937, s.25; PHA, 1961, ss.5(1), 11(2), sch.1, Part III; WA, 1973, s.14(7))

A local authority (i.e., a district council or London borough council) is required to reject plans of a building or its extension which are deposited with the local authority under building regulations and it is proposed to erect the building or extension over a sewer or drain shown on the map of sewers (see para.21), unless satisfied that in the particular case the authority may properly consent to the proposed erection either unconditionally or subject to special requirements. Any question between the local authority and the person by or on behalf of whom the plans have been deposited as to whether the site is over a sewer or drain or whether and upon what conditions the consent ought to be given may be determined by the magistrates court. A water authority may give directions to the local authority as to the manner in which the local authority is to exercise its functions under this provision.

16. CERTAIN MATTERS NOT TO BE PASSED INTO PUBLIC SEWERS

(PHA, 1936, s.27; CPA, 1974, s.99, sch.2, para.7)

A person who throws, empties or turns, or suffers or permits to be thrown or emptied, or to pass into a public sewer, or into a drain or sewer communicating with a public sewer:—

(a) any matter likely to injure the sewer or drain, or to interfere with the free flow of its contents, or to affect prejudically the treatment and disposal of its contents; or

(b) any chemical refuse or waste steam, or any liquid of a temperature higher than 110° Fahrenheit, being refuse or steam which when so heated, is, either alone or in combination with the contents of the sewer or drain, dangerous, or the cause of a nuisance, or prejudicial to health; or

(c) any petroleum spirit or carbide of calcium;
is liable (i) on summary conviction to a fine not exceeding £400 and to a further fine not exceeding £50 for each day on which the offence continues after conviction therefor; (ii) on conviction on indictment to imprisonment for a term not exceeding two years or a fine or both.

Nothing in paras. (a) and (b) above applies to trade effluents which may lawfully be passed into a public sewer by virtue of the Public Health (Drainage of Trade Premises) Act, 1937.

For the purpose of exercising powers or duties conferred by this provision no account is to be taken of any radioactivity possessed by any substance or article or by any part of any premises (Radioactive Substances Act, 1960, s.9(1)(2), sch.1, Part I, para.3) – this means that any effects from radioactivity are not the concern of the water authority and are dealt with by the Health and Safety Executive.

Definitions "Prejudicial to health" means injurious, or likely to cause injury to health (PHA, 1936, s.343(1))

"Petroleum spirit" means any such (a) crude petroleum, (b) oil made from petroleum, or from coal, shale, peat or other bituminous substances, or (c) product of petroleum or mixture containing petroleum, as, when tested in the manner prescribed by or under the Petroleum (Consolidation) Act, 1928, gives off an inflammable vapour at a temperature of less than 73° Fahrenheit (PHA, 1936, s.27(3)).

17. HOLDING LAND FOR TREATING SEWAGE

(PHA, 1936, s.29)

A water authority who holds land for treating the contents of its sewers may itself manage it in such manner as it thinks fit, or may lease it up to 21 years subject to provisions in the lease for ensuring the effective disposal without creating a nuisance of all sewage brought to the land.

18. PURIFICATION OF SEWAGE EFFLUENT

(PHA, 1936, s.30)

Nothing in Part II of the PHA, 1936, authorizes a water authority to construct or use a public or other sewer, or a drain or outfall for the purpose of conveying foul water into a natural or artificial stream, watercourse, canal, pond or lake until the water has been so treated as not to affect injuriously the purity and quality of water in the stream, watercourse, etc.

It has been held (in *Durrant v Branksome UDC* [1897] 2 Ch. 291) that effluent can be discharged into a watercourse provided the purity and quality of the water is unaffected, and that water from surface water sewers may be discharged to a stream even though it contains silt and sand. Where a water authority discharges sewage in contravention of this section the Attorney-

General may institute civil proceedings to restrain the authority where there is an infringement of statutory provisions (*A.-G. v Dorchester Corpn.* [1905] 69 J.P. 363). It is not necessary to show that the river has deteriorated in general, but that something has been added to the water which detracts from the purity and quality of the water at the point where the offending discharge entered the stream (*A.-G. v Birmingham Tame & Rea District Drainage Board* [1908] 2 Ch. 551).

A riparian proprietor (i.e. the owner or occupier of land on a river bank) has a right at common law to have the water of the stream come to him in its natural state, in flow, quantity and quality, and if unpurified sewage is passed to the stream and pollution results, he may bring an action for damages against the infringer without proving actual damage, unless the infringer has gained a right to pollute by long enjoyment or grant, and he may also obtain an injunction to prevent a continuance of the injury (*Jones v Llanrwst UDC* [1911] 1 Ch. 393). An action can also be brought by a non-riparian proprietor, namely by the owner of a fishery where the fish are driven away by foul drainage, or by a person whose well becomes contaminated.

In an action for pollution the plantiff may claim an injunction and damages based on nuisance and trespass or for negligence, or on the principle of *Rylands v Fletcher* (that the defendant is absolutely liable for any escape (e.g.) of a polluting discharge as a trespass) or for breach of statutory powers.

19. OTHER STATUTORY RESTRAINTS ON FOUL DISCHARGES

In addition to the prohibition on the discharge of foul sewage imposed under para.18, there exist other statutory restrictions on the discharge of sewage.

It is an offence under s.2(1)(a) of the Rivers (Prevention of Pollution) Act, 1951, for a person to cause or knowingly permit to enter a stream any poisonous, noxious or polluting matter, and for the purposes of that provision a water authority is deemed to cause or knowingly permit to enter a stream poisonous, noxious or polluting matter which passes into the stream from a sewer or sewage disposal works vested in it, where either the authority was bound to receive the matter into the sewer or works (i.e. under s.34 of the PHA, 1936 – see para.22), or it consented to do so unconditionally or subject to conditions which were observed (i.e. under the Public Health (Drainage of Trade Premises) Act, 1937, as amended – where the water authority grant consent following the service of a trade effluent notice). On conviction of an offence under s.2 of the 1951 Act the maximum penalty is a fine of £100 in the magistrates' court, or £200 in the Crown court.

It is also unlawful to make a discharge of sewage (or trade) effluent to a stream without the consent of the water authority, and in the case of discharges to be made by the water authority itself consent must be granted or deemed to be granted by the Secretary of State. This statement requires expanding because, strictly speaking, s.7 of the Rivers (Prevention of

Pollution) Act, 1951, refers to consent being required before a new or altered outlet for the discharge of trade or sewage effluent to a stream is brought into use (the outlet being one constructed after October 1, 1951) or a new discharge of trade or sewage effluent (i.e. a discharge not being one substantially a continuation of a previous discharge made within the previous 12 months) is begun to be made to a stream. Whilst the 1951 Act took care of effluents discharged or to be discharged to a stream after that Act became operative, it needed another statute, the Rivers (Prevention of Pollution) Act, 1961, to look after existing discharges commenced before 1951. Accordingly, s.1 of the 1961 Act provides that the making of a discharge of trade or sewage effluent to a stream, being a discharge which is not controlled under s.7 of the 1951 Act, requires the consent of the water authority.

Conditions relating to the nature or composition, temperature, volume and rate of discharge of the effluent and facilities for taking samples of the effluent may be reasonably imposed by the water authority in granting its consent, subject to a right of appeal to the Secretary of State. It is by means of imposing such conditions that the water authority obtains an important degree of control over the quality and quantity of sewage effluents.

Sometimes it is convenient and appropriate for sewage effluent to be discharged into a nearby aquifer, in which case it might be noted that trade or sewage effluent cannot be discharged into any underground strata within a water authority area by means of a well, borehole or pipe without the consent of the authority (see s.72 of the Water Resources Act, 1963).

It is also an offence under s.4 of the Salmon and Freshwater Fisheries Act, 1975, to cause or knowingly permit any liquid or solid matter to enter waters containing fish, or the tributaries thereof, to such an extent as to cause the waters to be poisonous or injurious to fish or their spawning grounds, spawn or food.

20. NUISANCE NOT TO BE CREATED

(PHA, 1936, s.31)

The water authority must so discharge its functions under ss.14 to 30 of the PHA, 1936, so as not to cause a nuisance.

There have been a number of decisions by the courts in this connextion affecting local authorities which would apply equally to water authorities in respect of transferred functions. A local authority which acted properly in exercising its statutory powers and had not been negligent was held not to be liable for the escape of sewage from overloaded sewers maintained by the authority (*Smeaton v Ilford Corpn.* [1954] 1 All E.R. 923). But an authority which exercise its powers negligently so as to cause a nuisance and injury can be sued by the injured party (*Hanley v Edinbugh Corpn.* [1913] A.C. 488). So far as nuisance is concerned, it now appears to be immaterial whether a local authority has or has not taken any steps which have lead to the nuisance being created, that is to say, an authority may be liable for misfeasance (e.g.

where an authority is negligent in the construction of a sewer, or directs sewage from one sewer to another which is already surcharged) and non-feasance (where an authority omits to enlarge a sewer to cope with an enlarging population) (*Pride of Derby & Derbyshire Angling Assocn Ltd. v British Celanese Ltd.* [1953] 1 All E.R. 179).

21. LOCAL AUTHORITY TO KEEP MAP OF PUBLIC SEWERS

(PHA, 1936, s.32; WA, 1973, s.40(2), sch.8, para.36)

Every local authority (i.e. a district council or London borough council) is required to keep deposited at its offices for inspection by all persons at reasonable hours free of charge a map showing and distinguishing all sewers and drains within its district being (1) public sewers, (2) sewers for which a vesting declaration has been made (see para.6) but which has not yet taken effect, (3) sewers or drains for which an agreement to make a vesting declaration in the future has been entered into (see para.7). The map must also show if any public sewer has been reserved for foul water only or for surface water only. The above provision continues to apply throughout Greater London and the council of every inner London borough has to supply a copy of the map to the Greater London Council.

22. RIGHTS OF OWNERS AND OCCUPIERS TO DISCHARGE INTO PUBLIC SEWERS

(PHA, 1936, s.34; WA, 1973, ss.14(4), 40(2), sch.8, para.37; CPA, 1974, s.99, sch.2, para.8)

As a general rule the owner or occupier of any premises or the owner of a private sewer is entitled to have his drain or sewer made to communicate with the public sewers of the water authority and thereby to discharge foul water and surface water from such premises or private sewer.

There are certain exceptions from this rule, thus, no one can discharge directly or indirectly into a public sewer:—

(a) any liquid from a factory (other than domestic sewage or surface or storm water) or any liquid from a manufacturing process, but despite this prohibition the owner or occupier of trade premises within the area of a water authority may, with the consent of the authority, discharge into the authority's public sewers any trade effluent from such premises under the provisions of the Public Health (Drainage of Trade Premises) Act, 1937, as amended;

(b) any liquid or other matter the discharge of which into public sewers is prohibited by or under any enactment (including an enactment under the PHA, 1936).

Also, where separate public sewers are provided for foul water and for surface water, a person cannot discharge directly or indirectly (1) foul water into a sewer provided for surface water, or (2) except with the water authority's approval, surface water into a sewer provided for foul water.

Nor may a person have his drains or sewer made to connect directly with a storm-water overflow sewer.

Subject to the code for breaking open streets (as contained in the PHA, 1936, ss.279, 282, the Public Utilities Street Works Act, 1950, and the Water Act, 1945, sch.3, Part VI), the owner or occupier of any premises is entitled to break up any street for exercising his rights as above and for examining, repairing and renewing drains or private sewers draining his premises into a public sewer. The owner or occupier must give notice of his proposals to the water authority, who within 21 days after receipt of this may by notice refuse to permit the communication to be made if it appears to the authority that the mode of construction or condition of the drain or sewer is such that the making of the communication would be prejudicial to its sewerage system and for examining the mode of construction and condition of the drain or sewer the authority may require it to be laid open for inspection. Any question as to the reasonableness of any such requirement or of the authority's refusal to permit the communication to be made may be decided by the magistrates' court on the application of the person proposing to make the communication.

Where a person proposes to make a communication between a drain or sewer and a main drain in Greater London, the water authority may refuse to permit the communication on such grounds as it thinks fit and no application may be made to the magistrates court in respect of any such refusal by the water authority. In this connection, "main sewer" means a public sewer used for the general reception of sewage from other public sewers and not substantially used for the reception of sewage from private sewers and drains.

Where the water authority does not elect to make the communication itself under s.36 of the PHA, 1936 (see para.23), the person making it must give prior reasonable notice to any person directed by the water authority to superintend the execution of the work and afford him all reasonable facilities for this.

Anyone who causes a drain or sewer to communicate with a public sewer without complying with, or in contravention of, any of the above provisions, or before the expiry of the period of 21 days (within which the water authority may refuse permission for the communication), is liable to a fine not exceeding £200, and whether proceedings are taken or not, the water authority may close any contravening communication made and recover its reasonable expenses in doing so from the offender.

23. WATER AUTHORITY MAY MAKE COMMUNICATIONS WITH PUBLIC SEWERS

(PHA, 1936, s.36; Local Government (Miscellaneous Provisions) Act, 1953, s.13; CPA, 1974, s.99, sch.2, para.9)

Where notice is given (under para.22) to a water authority that a person proposes to have his drains or sewer made to communicate with a public sewer of the authority, the authority may within 14 days after receipt of the notice notify the person that it intends itself to make the communication, and the authority has all such rights in respect of the making of the communication as the person desiring it to be made would have. If the person proceeds to make the communication after he has been so notified, he is liable to a fine not exceeding £200. The authority is not obliged to proceed with the work until the cost as estimated by its surveyor has been paid or security for payment given. However, the authority may decide to bear the whole or part of the cost itself.

24. WATER AUTHORITY MAY ALTER DRAINAGE SYSTEM OF PREMISES

(PHA, 1936, s.42)

Where any premises having a drain or sewer (i.e. a private sewer – the alteration, etc of public sewers comes under para.12) communicating with a public sewer or cesspool, but such drainage system, though sufficient for the effectual drainage of the premises, is not adapted to the general drainage system of the district, or is in the water authority's opinion otherwise objectionable, the authority may at its own expense and on condition that it first provides in a position equally convenient to the owner of the premises a drain or sewer equally effectual for the drainage thereof and communicating with a public sewer, close the existing drain or sewer and fill up any cesspool and do any necessary work. The water authority has to give notice of its proposal to the owner, who if aggrieved either as regards the position or sufficiency of the proposed drain or sewer, may appeal to the magistrates' court.

25. SEWERING OF HIGHWAY

(PHA, 1961, ss.12-14, sch.2)

If a water authority resolves to construct a sewer in a street or part thereof being a highway maintainable at the public expense and the resolution includes a declaration that the construction of the sewer will in the opinion of the water authority increase the value of premises fronting the street, then the authority may recover payments from the owners of the premises for the cost of the sewer. The authority may remit part of the payments on specified grounds. The procedure for recovering payments and the formula for the amount of repayment are set out in the PHA, 1961, s.12 and sch.2.

Under s.13 of the PHA, 1961, the water authority may also by resolution decide to recover from the owners of premises in a street payments for the construction of a sewer in land which after the construction of the sewer has been laid out as a street or part thereof.

The water authority may, under s.14 of the PHA, 1961, apply to the magistrates' court for an order in cases where an owner of premises attempts to evade liability to make payments under ss.12 or 13 by a transfer of land.

The expenses of sewering private streets may be put on the frontagers – see the Highways Act, 1959, Part IX referring to private street works codes for making up and sewering of private streets.

26. COMPENSATION

(PHA, 1936, s.278)

A water authority must make full compensation to any person who has sustained damage by reason of the authority exercising any of its powers under the PHA, 1936, relating to a matter which he has not himself been in default. Any dispute about the fact of damage or the amount of compensation is determined by arbitration and when the claim does not exceed £50 all questions as to the fact of damage, liability and amount may be determined on the application of either party before the magistrates' court. Where a person claims compensation for damage sustained from the construction of a sewer in or over his land, the tribunal must determine the amount by which the value to the claimant has been enhanced by the cost of the sewer and this amount may be set off against the amount of compensation awarded.

27. POWER TO ENTER PREMISES

(PHA, 1936, ss.287, 288)

On producing a duly authenticated document showing his authority, an authorized officer of a water authority or local authority is entitled to enter upon any premises at reasonable hours (subject to giving 24 hours prior notice to the occupier of premises which is not a factory, workshop or workplace) to ascertain whether there has been any contravention of the provisions of the PHA, 1936, or whether circumstances exist which would authorize or require the authority to take any action or execute any work under the Act, or to take any action or execute works as required by the Act, or also generally for performing its functions under the Act. In the event of actual or apprehended refusal of admission to premises, or unoccupied premises, or temporary absence of the occupier, or case of urgency, or where application for admission would defeat the object of entry, and there is reasonable ground for entry for any such purpose, a justice of the peace may authorize entry by warrant. Wilful obstruction of entry to a person renders the offender liable to a penalty not exceeding £5 and a further penalty of £5 for each day on which the offence continues after conviction therefor.

28. WORKS AFFECTING WATER RIGHTS

(PHA, 1936, s.331)

Nothing in the PHA, 1936, authorizes a water authority injuriously to affect a reservoir, canal, watercourse, river or stream, or any feeder thereof, or the supply, quality or fall of water contained in, or in any feeder of, any reservoir, canal, etc without the consent of the person who would, if the Act had not been passed, have been entitled by law to prevent, or be relieved against, the injurious affectation of, or of the supply, quality or fall of water contained in the reservoir, canal, etc.